Everyday Physical Science Experiments With
Light

Amy French Merrill

Rosen Classroom Books & Materials™
New York

For Kelli

Some of the experiments in this book are designed for a child to do together with an adult.

Published in 2006 by The Rosen Publishing Group, Inc.
29 East 21st Street, New York, NY 10010

Copyright © 2006 by The Rosen Publishing Group, Inc.

All rights reserved. No part of this book may be reproduced in any form without permission in writing from the publisher except by a reviewer.

First Edition

Book Design: Michael Caroleo, Michael de Guzman, Nick Sciacca
Project Editor: Frances E. Ruffin

Photo Credits: Cover (lamp), pp. 5 (lamp), 17 (eyeglasses and magnifying glass) © Eyewire; cover (flashlight with shadow, girl), pp. 8, 9, 13–15, 18–21 by Adriana Skura; pp. 5 (candle), 7 by Michael Flynn; p. 11 (submarine) © Yogi, Inc./Corbis; p. 11 (inset) © Steve Kaufman/Corbis; p. 17 (microscope) © Corbis.

Merrill, Amy French.
  Everyday physical science experiments with light / Amy French Merrill.
    p. cm. — (The Tony Stead nonfiction independent reading collection)
  Includes bibliographical references and index.
  ISBN 1-4042-5679-2
  1. Light—Experiments-Juvenile literature. 2. Sound—Experiments—Juvenile literature. [1. Light—Experiments. 2. Sound—Experiments. 3. Experiments.] I. Title. II. Series.
  QC360.M467 2006
  535'.078—dc22

2005010028

Manufactured in the United States of America

CPSIA Compliance Information: Batch #WR014140RC: For further information contact Rosen Publishing, New York, New York at 1-800-237-9932.

Contents

1. What Is Light? — 4
2. Transparent or Opaque? — 6
3. Fun with Shadows — 8
4. Reflecting Light — 10
5. Starting Your Periscope — 12
6. Finishing Your Periscope — 14
7. Bending Light — 16
8. A Broken Spoon? — 18
9. The Colors of Light — 20
10. So Much Light — 22
    Glossary — 23
    Index — 24
    Web Sites — 24

What Is Light?

Light is a form of **energy**. It can come from nature or be man-made. If you look at the beam of a flashlight in a dark room, you can see that rays of light travel in straight lines. If you hold an object in front of the flashlight, the rays pass through, bounce off, or are **absorbed** by the object, depending on what the object is made of.

We need light in different forms. ▶
It helps us get important information about the world around us.

Transparent or Opaque?

Find objects that are made of different **materials**, such as a clear glass, a book, and a piece of cardboard. Hold them in front of a lit flashlight one at a time. What do you see? Materials that let light pass through, like glass, are called **transparent**. Materials that stop light from passing through are called **opaque**.

Can you tell which objects are opaque and which objects are transparent? ▶

You Will Need:

book, cardboard, flashlight, glass

Fun with Shadows

A shadow forms when light cannot pass through an opaque object. To make a shadow, draw the outline of an object on a sheet of construction paper. Cut it out and tape it onto a stick. Facing a blank wall, hold your cutout in front of a flashlight. The cutout blocks the light and casts a shadow on the wall.

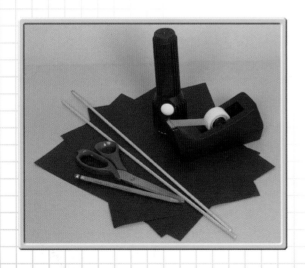

You Will Need:

flashlight, construction paper, pencil, scissors, tape, wooden stick

You can make more cutouts and have a shadow puppet show. ▼

Keep the light on the cutout but pull the flashlight farther away from it. The shadow on the wall will grow larger. ▼

Reflecting Light

How important is light? Without light we cannot see. We see because light **reflects**, or bounces, off objects. Different surfaces reflect light in different ways.

One way to understand reflected light is by using a **periscope**. A periscope uses mirrors to help the viewer see over or around something that is in the way.

Without periscopes, people in submarines—like the man using a periscope in the upper left—could not see above water. ▶

Starting Your Periscope

To make your own periscope, follow these steps:

1. Use a ruler to draw two **diagonal** lines in the same places on opposite sides of a cardboard milk carton.
2. Have an adult help you cut **slots** along the lines you drew.
3. Push a mirror into the top slot so that the mirror side faces down.
4. Push a mirror into the bottom slot so that the mirror side faces up.

You Will Need:

milk carton, ruler, pen, scissors,
two mirrors (each a bit wider than the carton)

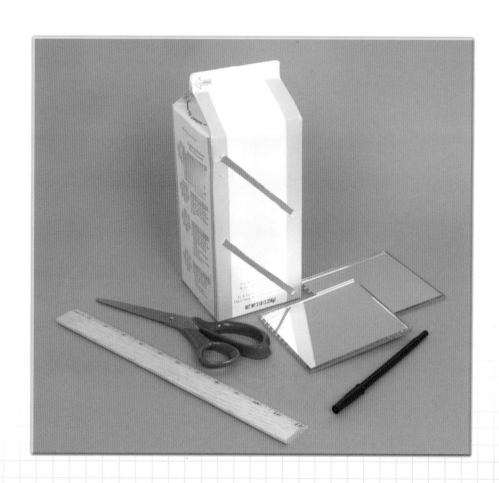

Finishing Your Periscope

5. On the front of the carton, cut out a square opening between the two mirrors. The opening should be close to the top mirror.
6. Punch a small hole with a pen in the back of the carton between the two mirrors. The hole should be close to the bottom mirror.
7. Stand near a corner and hold the carton on its side so that only the square opening sticks out from the corner.
8. Peek into the small hole.

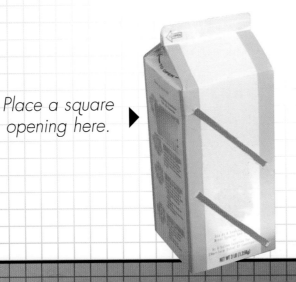

Place a square opening here.

Punch a small hole here.

You should be able to see around the corner!
▼

Bending Light

Light rays change direction when they move from one transparent material to another. For example, light rays bend when they pass from air to water or from water to air. This bending is called **refraction**. Refraction is very useful. Curved pieces of glass that refract light, called lenses, can correct a person's vision.

Lenses are used in eyeglasses, magnifying glasses, and microscopes. ▶

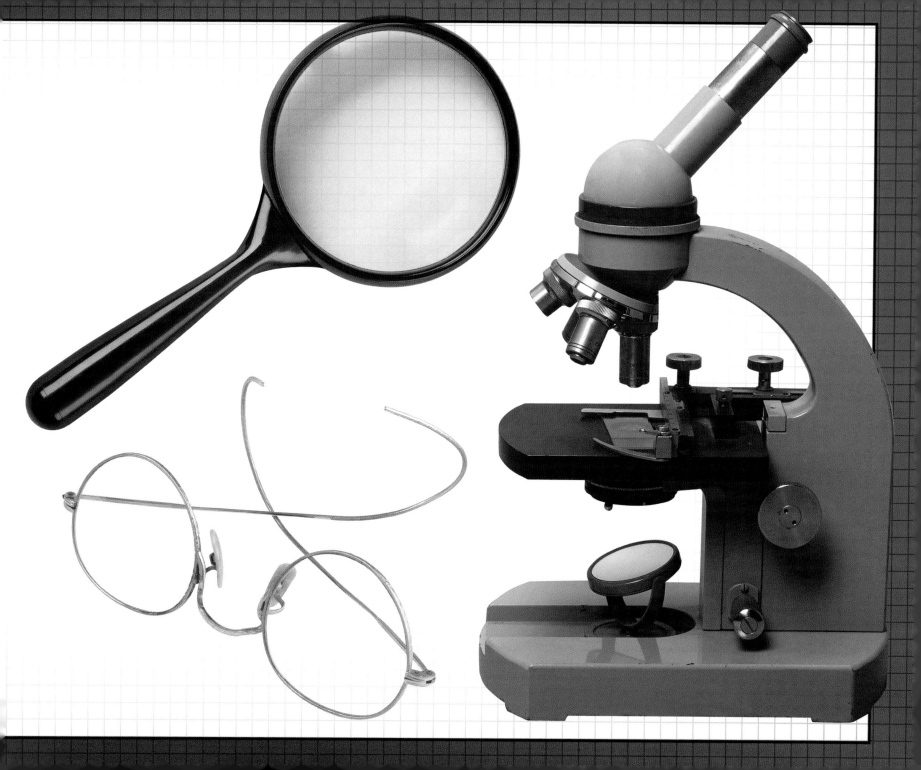

A Broken Spoon?

This experiment shows refraction at work. Rest a spoon at an angle in a glass. Pour water into the glass. Stand back and look at the glass. The spoon appears to break into two sections. Why does the spoon look broken? Light rays bend as they pass from water into air. The bending light rays make the spoon look as though it is broken.

You Will Need:

glass, water, spoon

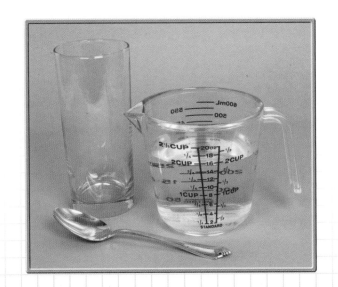

18

Light rays move more slowly through water. The change in speed from air to water causes the rays to bend.
▼

The Colors of Light

When sunlight is refracted as it passes through raindrops in the air, a rainbow forms. This shows us that light is made of colors.

Try this experiment. Cut a circle from a piece of white cardboard. Divide the circle evenly into seven sections. Paint or color each section with one of the colors of the rainbow. Make a tiny hole in the center of the circle and fit a toothpick into it. Spin the wheel. What happens to the colors?

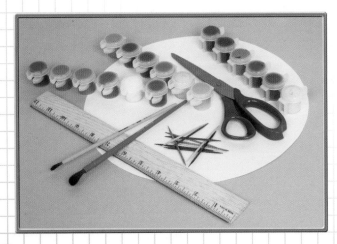

You Will Need:

white cardboard, scissors, ruler, paints or markers (red, orange, yellow, green, blue, indigo, violet), paintbrushes (if paint is used), toothpick

Be sure to paint or color each section in this order: red, orange, yellow, green, blue, indigo, and violet.

When you spin the wheel, your eyes and brain cannot focus on the individual colors. The colors blend together into white.

So Much Light

How do people use light? Photographs—pictures created with light—are lasting images of people, places, and events in history. Traffic lights help direct cars, trucks, and buses. Powerful beams of light, called lasers, read special codes of information on items in stores. We use light every day in so many ways!

Glossary

absorb (uhb-SORB) To take in moisture, heat, or light.

diagonal (dy-AA-guh-nuhl) A straight line that cuts across in a sloping direction.

energy (EH-nuhr-jee) The power to work or act. Light and heat are forms of energy.

material (muh-TEER-ee-uhl) What something is made of.

opaque (oh-PAYK) Blocking light so that it cannot pass through.

periscope (PEHR-uh-skohp) A tool that uses mirrors or lenses to see around or over objects.

reflect (rih-FLEKT) To throw back light.

refraction (rih-FRAK-shun) The act of bending rays of light.

slot (SLAHT) A narrow opening.

transparent (trans-PEHR-uhnt) Allowing light to pass through.

Index

A
absorbed, 4

B
bend(ing), 16, 18
bounce(s) off, 4, 10

C
colors, 20

E
energy, 4

L
lasers, 22
lenses, 16

M
mirror(s), 10, 12, 14

O
opaque, 6, 8

P
pass(ing) through, 4, 6, 8
periscope, 10, 12
photographs, 22

R
rainbow, 20
rays, 4, 16, 18
reflect(s), 10
refract(ed), 16, 20
refraction, 16, 18

S
shadow, 8

T
traffic lights, 22
transparent, 6, 16

Web Sites

Due to the changing nature of Internet links, the Rosen Publishing Group, Inc., has developed an online list of Web sites related to the subject of this book. This site is updated regularly. Please use this link to access the list: http://www.rcbmlinks.com/tsirc/light/